Feeling the Vibrations of the Universe

Jim Eerie

Pichu Press

*A catalogue record for this book is available
from the British Library*

ISBN: 978-1-907962-45-5

Published by Pichu Press

Reading, England

Challenge

Question

Ponder

For Karen

Contents

Preface

The purpose of this book is to open you up to possibilities that you might not have considered. These possibilities, when considered, could potentially bear great fruits. Your day-to-day interactions with the world might be based on a set of unquestioned assumptions. These assumptions are likely to result in a restricted view of you, of the universe, and of the relationship between you and the universe.

It is my aim to expose the unquestioned assumptions; to reveal the possibilities that exist. It is only when one realises that the possibilities exist that one can embrace them.

Introduction

In this book I wish to explore the idea that the universe is made of vibrations, that these vibrations feel a certain way, that the vibrational flows of the universe are deeply interconnected, and that you can become aware of feeling vibrations which exist outside of your body.

I believe that there are great benefits to realising that this is the nature of the universe. You will be able to see yourself as part of a larger whole and be able to connect deeply with that larger whole. You may even reach the stage in which wherever you are, wherever you go, you seek to attempt to feel the vibrations that exist in this particular location. Good luck!

Feeling the Vibrations of the Universe

I am aware that what I am about to say is likely to strike you in one of two ways. You might think *this is very strange, how could this possibly be right,* or, contrarily, you might think *of course, this is obvious.* I hope that I have something of interest to say to both of these potential audiences.

What I am about to say concerns the fundamental nature of the universe – vibrations – and how humans can become attuned so that they feel these vibrations. The magnificent vibrations! Feeling them – connecting with them; this can be a great source of joy, a great source of fulfilment!

You might find such a notion to be exceedingly odd. If so, I presume that this is *not* because you find the idea that the universe is composed of vibrations to be exceedingly odd. Rather, I presume that it is because you find the idea that you can feel these vibrations – vibrations which exist outside of your body – to be exceedingly odd.

Why do I presume such a thing? Well, you might think that the fundamental nature of the universe is unknowable – fundamentally mysterious. If so, you would be correct. However, we can draw conclusions about the likely nature of the universe – or, more safely, given the constraints on our knowledge, we can say *as far as we know this is the most likely nature of the universe.* Even more weakly, we can say, given the constraints on our knowledge, *this is a very plausible candidate for the fundamental nature of the universe.*

Given our knowledge, the view that vibrations are the fundamental nature of the universe is very plausible. More than this, as far as we know this seems (possibly arguably) to be the most likely fundamental nature of the universe. This view of the universe is exemplified by leading scientific theories such as String Theory – according to which tiny vibrating loops are the fundamental constituents of the universe. So, this is why I believe that you don't find the idea that vibrations could be the fundamental nature of the universe to be exceedingly odd. Indeed, rather than being exceedingly odd, it is (given the constraints on our knowledge) possibly the most likely nature of the universe.

So, I presume that if you find the idea that humans can feel the vibrations of the universe to be exceedingly odd, that this is because you doubt that you have the ability to feel things (to 'tune into' things) which exist outside of your body. If you have this doubt it is quite understandable; you will probably be in the majority. I would think that most people believe that they can only feel, or be affected by, that which is inside their body.

What I have just said is a bit too simplistic. For I take it that the overwhelming majority of people do believe that they can be affected by things which exist outside of their body. For example, imagine that you are about to walk across a road and have taken the first step onto the road when you suddenly hear a loud 'horn' noise. You look to your left and see a car speeding towards you. You step backwards. You have clearly been affected by things which exist outside of your body – the pressing of the horn, the movement of the car.

So, it is obvious that people *are* affected by what happens in the world outside of their bodies. The question of importance is how extensive such affects are. On one end of the scale you might think that such affects are limited to the human senses – sight, hearing, taste, smell, touch. We have seen that *one can be affected by* the horn blasting and the movement of a car which exists in the world outside of one's body.

However, in the above example, there is more to 'being affected by' than simply taking a step backwards. It is obvious that the feelings within one's body can be affected by the things in the world outside of one's body. The blasting of the horn, the movement of the car, can result in an increase in one's pulse and to 'feelings of stress' to start streaming through one's body.

So, what this means is that, if the universe is comprised of vibrations, that the vibrations that exist in the universe outside of your body can affect the vibrations of the universe that exist within your body. The movement of the car, and the blasting of the horn, are 'vibrational events' which lead to 'vibrational events' in your body.

It is possible that you wouldn't have thought about this event in these terms before. When one thinks in these terms it sheds a partial light on the interconnected nature of the universe. We have considered how vibrations that exist outside of your body can affect the vibrations within your body. The effects we have considered so far have been mediated by one's senses. If one had not heard or seen the external events that were 'horn blasting' and 'car moving' then one's internal vibrations – we have assumed, so far – would not have been affected.

There are clearly other ways in which the vibrations within one's body could be affected by the vibrations which exist outside of one's body. For one thing, there are good reasons to believe that vibrations resonate with other. What does this mean? Let me attempt to paint a picture in your imagination!

Focus solely on your body – your beautiful vibrating body! How wonderful it is! Full of vibrations, it is vibrating in a multitude of different ways. A bodily symphony is formed as the various vibrations interact with each other. Initially there can be inharmonious vibrations resulting in a discord. Slowly the vibrations intermingle, they affect each other – slowly but surely – affect, modify, change – bring into closer harmony. The bodily vibrations envelop each other until a bodily symphony is forged – the vibration pattern of the body as a whole. There is a dominant vibration pattern of the body which is attained when the dominant vibration patterns encapsulate the lesser vibration patterns – transforming and embedding them within their dominant pattern.

Vibrate, vibrate my body – reveal my essence through your vibrations

Now, focus solely on the universe *that is not* your body. There is an immense array of diverse vibration patterns which intermingle as they flow through the universe on their cosmic journeys. Flow, flow, flow – modify the vibrations that you inter-mingle with as you flow. Seek to transform the entire universe to the vibrational essence that is you (in this context 'you' = all the vibrations in the universe).

How wonderful! Imagine all the flows of the universe (all of the universe) as they flow and intermingle and gently affect each other. As the more pervasive flows gently modify the minority vibration patterns a dominant cosmic vibrational pattern is forged.

Vibrate universe, vibrate; reveal your inner essence

Imagine a ray of sunlight as it flies away from the Sun on its journey to the Earth. This ray is a particular vibration pattern that is affecting and modifying its surroundings as it travels. Forever vibrating. Forever transforming. Go my ray of sunlight, go, vibrate, transform, change that which is not you to you! Keep on travelling; keep on trans-forming.

Imagine the vibrations that exist in clouds as they glide through the sky above. As the vibrations change, the clouds change to rain. Vibrate my clouds, rain my clouds!

Vibrate as you need to – let the rains pour!

Imagine the vibrations that are a flock of birds flying through the sky, a stream flowing on its inevitable course as its vibrations interact with the vibrations of the surrounding ground. Imagine the vibrations of the fish as they swim though the stream. Imagine the vibrations of the fisherman as he stands by the stream, the vibrations of his fishing rod, the vibrations of his bait, and the change in vibrations in a fish when it gets caught on the hook of the fishing rod. Vibrate my fish, vibrate!

Now, let us bring these two exercises of the imagination together. Let us imagine the vibrational flows of the universe as they interact with the vibrations of your body. How wonderful! Imagine that you are standing on the Earth. Your body has a particular vibrational essence. It has a dominant vibrational essence, and also a variety of lesser vibrational patterns which are 'under the wings' of the dominant essence. Imagine the vibrational pattern of the ray of sunlight as it journeys from the Sun. Imagine its particular vibrational essence.

Imagine it travelling, travelling, vibrating, transforming. Imagine the moment this ray strikes your body. Wow! Its vibrational essence is now inside you, transforming you!

Transform me sunlight; bring your light to me; change my essence

The vibrational essence of the ray of sunlight will be intermingling with the vibrational essence of your body. The vibrations of your body will be subtly changed forever. A new dominant vibrational essence may arise in your body, or the vibrational essence of your body may overpower that of the ray resulting in no change to your dominant vibrational essence.

Imagine the vibrational pattern of a raindrop as it journeys to the Earth. Imagine the raindrop falling onto your head. Imagine its vibrational pattern interacting with your bodily vibrations. Wonderful! Modify raindrop modify, cleanse my vibrations, purify, purify!

Let us recap. We have been exploring how the vibrations that exist in the universe around one can change the vibrations in all parts of one's body by interacting with them. If a raindrop falls onto one's finger it will intermingle with the vibrational patterns in one's finger; these new vibrational patterns will then spread out from the finger to surrounding areas in a 'ripple' effect. The same thing will happen if a raindrop hits your foot; only the starting point in your body will be different.

So, when a raindrop hits your foot there will be an intermingling between the vibration patterns of raindrop and foot. Either the previous vibration pattern in your foot will remain dominant and there will be no noticeable changes in the vibration pattern, or their will be noticeable changes in the vibration pattern. If there are noticeable changes it is possible that you might become aware of these changes.

This is an important point to appreciate. You are often aware of the vibration patterns that exist in various parts of your body. You might sometimes refer to a particular pattern of vibrations in your body as a "pain in the arm" or "pins and needles in the leg" or a "headache", and so on. We have seen that these vibration patterns not only exist in your body – they exist throughout the universe.

The universe is an interconnected flow of vibrations. You can become aware of the vibration patterns in your body. The question I wish to ask is: Can you become aware of the vibration patterns which exist outside of your body?! The answer to this question seemingly depends on the issue of whether the skin of your body forms an impermeable barrier between 'body' and 'non-body universe'. When one appreciates that the universe is an interconnected flow of vibrations it is hard to take seriously the idea that any parts of the universe are impermeable!

You might have trouble accepting this idea. Conventional wisdom still seems to dictate that such an impermeable barrier exists – dictates that you can only become aware of the feeling vibrations that are located in your body. I find the term 'feeling vibrations' useful as it makes it clear that we are not talking about become aware of vibrational events such as 'horn honking'; rather, *we are talking about become aware of the feelings which exist in the vibrations which are generated in the event which causes one to hear that a horn has been honked.*

Is this clear? Let me reword it. When a horn is honked a vibrational event has occurred in the location of the horn and vibrations spread out from the horn. These vibrations have feelings – all vibrations are feelings. When one hears the vibrational pattern that emanates from the horn one is hearing a sound, one is not hearing the feeling. There are two sides to the event – feeling vibrations and sound heard.

If your body is not an impermeable barrier then there is no reason why you should not be able to both hear the sound and feel the vibrations too. There are many reasons to believe that this is possible. There are a plethora of reports of people suddenly becoming overwhelmed with feelings of 'sickness'/'sadness'/'terror' and later finding out that a distantly-located loved-one died at that exact moment. Scientific experiments show that when certain parts of a human body are taken to a distant location they react to stress (= a particular vibration pattern) in the body that they were previously in.

And, of course, experiments in the realm of quantum physics imply that the vibration patterns of the entire universe affect each other.

The combination of the entanglement of the universe with the permeability of the body produces a wonderfully interconnected view of the universe!

The universe is an interconnected flow of entangled vibration patterns; these patterns flow through your body. The vibrations in any one part of the universe affect both the vibrations that they flow through and distantly located vibrations via entanglement. You can become aware of both the entangled flows in your body & the entangled flows which exist outside of your body!

What does this mean? It means that the feeling vibrations that are in your body are affected by things such as rays of sun, the amount of cloud cover, the flight of a bird, and the position of the moon. It also means that some of the feeling vibrations that you are aware of are not located in your body – they are located in the non-your-body universe. How wonderful!

You might not have realised this before. You might have assumed that all of the feeling vibrations that you have become aware of (fear, unease, tension, elation, tingling, ecstasy, and so on) are solely located within your body. This wouldn't be surprising. When we're born we are given a 'name' and this serves as a mechanism which acts to establish, and reinforce in our thinking, that we are essentially isolated from and separate from the non-your-name universe. It is 'Fred' versus the 'universe', or 'Mary' versus the 'universe'!

The vibrational flows of the universe do not care about our 'names', our conventions, our thoughts concerning where the feeling vibrations that we become aware of are located. They just keep entangling, vibrating, transforming and flowing!

If you get stuck in the conceptual framework according to which you are locked within your body, and have no access to the feeling vibrations which are in the non-your-body-universe, then you will not be able to fully connect with your surroundings. Whatever happens, whatever you become aware of, you will assume: *this is a state solely arising within my body.* You will be like a prisoner stuck inside their cell.

Escape your cell my friend – scale the prison walls. Realise that you can connect at a deep level with your surrounding universe. You can connect with, feel, become aware of, the feeling vibrations located in that which is not you. Why would you not want to do this?

Open your mind

Realise that your body is already open

Connect with the feeling vibrations of the universe

Books by the author:

Living in the Mesocosm   (2011)

Is Life a Concept?   (2011)

Why do People Walk on the Path?   (2011)

Feeling the Vibrations of the Universe   (2011)

Why do Humans Work so Hard?   (2011)

Effortless Achievement   (2011)

www.ingramcontent.com/pod-product-compliance
Lightning Source LLC
Chambersburg PA
CBHW060514210326
41520CB00015B/4219